Distribution and Synthesis of Alkaloids Among Plant kingdom

Authors

Dr. Khushal M. Kapadiya
Assistant Professor,
School of Science,
RK University,
Rajkot- 360020
Gujarat (India).

Dr. Ranjan C. Khunt
Associate Professor,
Department of Chemistry,
Saurashtra University,
Rajkot- 360005
Gujarat (India).

RK UNIVERSITY

Preface

The goal of this book to provide a comprehensive set of basic notes on chemistry of alkaloids, which will be suitable for undergraduate and post graduate students taking chemistry, chemistry related courses or course which involve chemistry of natural products. The book focus on core topics which are most likely to be common to those organic chemistry course which follow on from a foundation or introductory general chemistry of alkaloids.

The present book addresses itself of these objects. Every chapter includes primary introduction of particular alkaloids, their origin, isolation process, and physiology in plant, structure elucidation and their synthesis.

In the present scenario on drugs and their adverse effect on youth, to understanding this thing, here especially we include one chapter on Heroin.

It is hope that students will find this book useful in their studies and that once they have grasped what is the importance of alkaloids in life, as a drug.

We express our sincere gratitude to the faculty members, students and authority of Saurashtra University and RK University for their kind support to write this book.

We are very much grateful to Lulu Publication for the prompt publication of this book in very short time.

Author

Table of content

1.0   INTRODUCTION

The term alkaloid was first time projected by W. Meissner in 1819.

The term "alkaloid" (alkali-like) is commonly used to designate **basic** heterocyclic nitrogenous compounds of plant origin that are physiologically active.

Deviation from Definition:

➢ Basicity:

Some alkaloids are not basic e.g. Colchicine, Piperine, Quaternary alkaloids.

➢ Nitrogen:

It is also found that even nitrogen is not present in the ring. Such as Ephedrine, Colchicine, and Mescaline

➢ Origin of alkaloids: The main sources of alkaloids are plant kingdom. However some alkaloids are also found in Bacteria, Fungi, Insects, Frogs, and Animals.

New Definition:

From the above information, in past alkaloids was define as "physiologically active compounds of plant source in which at least one nitrogen atom forms a post of cyclic system". After some time, it is found that there are some optimal cases which are nit basic in nature such as Colchicine, Piperine, Quaternary alkaloids etc.

2.0 DISTRIBUTION AND OCCURRENCE

Plants are main sources of alkaloids; they are found only 10-15 % in all vascular plants. More than 2500 types of alkaloids are isolated from different parts of the plant such as roots, seeds, leaves in a various plant families.

In most of the cases a particular plant contains closely related alkaloids, i.e. nearly 20 alkaloids have been isolated from opium.

The alkaloid constitution is varies with the season, age and plant locality i.e. complicated alkaloids are generally occurring in one species of genus of a family.

In general alkaloids are basic in nature. So generally it is found as a salt form in plant acids. Some alkaloids are also found in free form, e.g. Nicotine & Nerceine.

3.0 DISTRIBUTION IN PLANT

- ➢ All Parts e.g. Datura

  Coca tree (Cocaine)

- ➢ Barks e.g. Cinchona(Quinine)

- ➢ Seeds e.g. Hemlock (Coniine)

- ➢ Roots e.g. Ipecac (Emetine)

- ➢ Fruits e.g. Black pepper (Piperidine)

- ➢ Leaves e.g. Tobacco (Nicotine: 2-8 %)

- ➢ Latex e.g. Opium (Morphine)

4.0   PHYSIOLOGY

1 Alkaloids are act as protective agents against insects and herbivores due to their toxicity.

2 They may be utilized as a source of energy in case of deficiency in carbon dioxide assimilation.

3 In certain cases, it comes as the final products of detoxification (waste products).

4 Alkaloids also work as a nitrogen source for protein synthesis and other fragments.

5 As a regulatory growth functions.

5.0   PHYSICOCHEMICAL PROPERTIES

Solid crystalline compound (exceptions are: Coniine and Nicotine)(It doesn't have Oxygen in their structure).

Colorless compound (exception are Berberine; yellow, Betaine; red).

Sharp melting Point because it's pure compound in crystal form.

Can be either 1^0, 2^0, 3^0 or 4^0 alkaloid.

Basicity depends on availability of lone pair of electrons:

1. Electron donating or electron withdrawing neighbors.

2. Type of hybridization.

3. Aromaticity.

6.0   CLASSIFICATIONS

When structures of most alkaloids are not known, first classification was developed **according to the plant genera** from which they are occur.

Hemlock: Coniine

Tobacco: Nicotine

Cinchona: Quinine, Cinconine

Opium: Papavarine, Morphine

Coca: Cocaine

Rauwolfia: Reserpine

Strychonus: Strychnine

Now in present alkaloids are mainly classified **according to the ring system** which is common in group of alkaloids.

- ➤ **Phenylalkylamines:** e.g. Ephedrine

- ➤ **Pyrolidine:** E.g. Hygrine

- ➤ **Pyridine - Pyrolidine:** e.g. Nicotine

- ➤ **Tropane:** e.g. Atropine

- ➤ **Quinoline:** e.g. Quinine, Cinconine and Quinidine

- ➤ **Isoquinoline:** e.g. Papavarine, Narcotine

- ➤ **Phenanthrene:** e.g. Morphine

- ➤ **Indole:** e.g. Ergometrine, Reserpine, Strychnine

- ➤ **Imidazole:** e.g. Pilocarpine

- ➤ **Purine:** e.g. Caffeine

- ➤ **Steroidal:** e.g. Solanum and Veratrum alkaloids

- ➤ **Terpenoid:** e.g. Taxol

One more class of alkaloids was also used in past **according to their pharmaceutical importance**.

For Note:

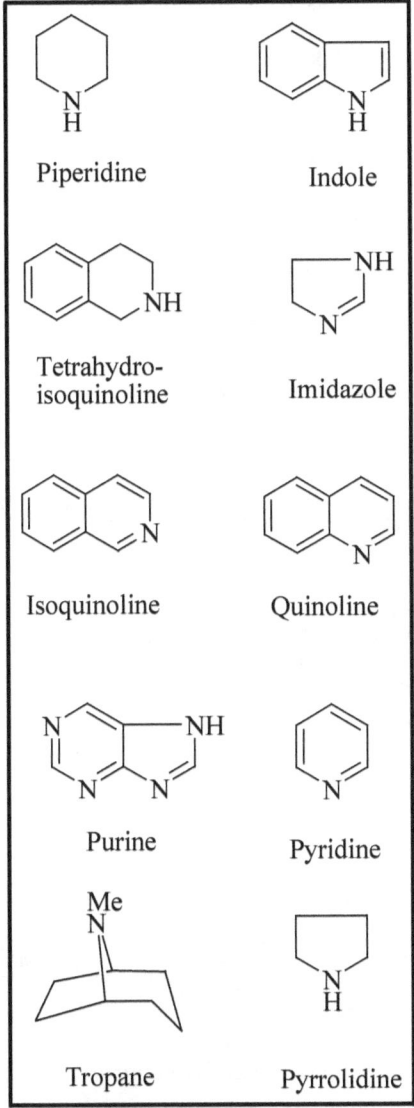

7.0   ISOLATION OF ALKALOIDS

Firstly presence of alkaloids was checked by treating plant extract with various alkaloidal reagents such as tannic acid, picric acid, perchloric acid, potassium mercuric iodide (Mayer's Reagent), which may form precipitates, turbidity or color change.

However, these alkaloids are not of great value, because the resulting compounds are not sufficiently soluble and the reagents precipitate other organic substance also.

A general procedure is explained by **manske** as below:

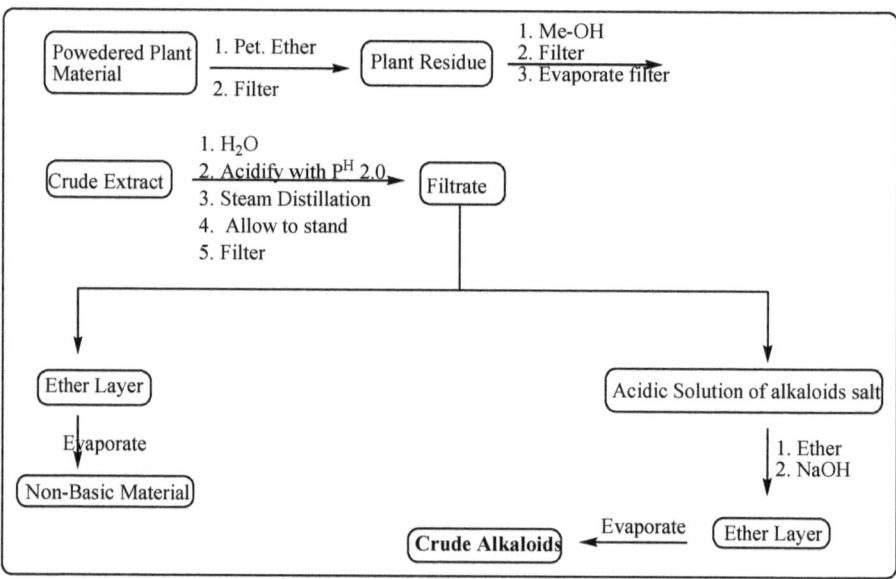

The pure form of single alkaloids can be separated by fractional distillation or other chromatographic techniques from alkaloids fraction.

8.0   QUALITATIVE ANALYSIS

Detection of alkaloid is carried out by following reagents:

Wagner's test (I_2/KI): Reddish brown precipitate,

Mayer's test ($KHgI_2$): Creamy precipitates with True alkaloid,

Hagger's test (Picric acid): Yellow precipitate with True alkaloid,

Dragendroff test (Potassium Bismuth Iodide): Reddish Brown precipitate,

Tannic acid solution test: different colored alkaloid precipitate.

9.0   GENERAL STRUCTURE ELUCIDATION OF ALKALOIDS

Step I: Molecular Formula

It can be determined by elemental analysis and mass spectrometry.

Step II: Optical Activity

It can be measured by specific rotation.

Step III: Cleavage

Alkaloid is cleaved in to simple fragments by hydrolysis with water and fragments are examined separately.

e. g. On hydrolysis of piperine, it gives two fragments Piperidine and piperic acid. It indicates that these two fragments linked with each other by amide linkage.

Step IV: Nature of Oxygen atom

Oxygenated functional group present in alkaloid can be identified by following reactions:

(A) Phenolic –OH:

Number of phenolic –OH groups is identified by acetylation.

$$R\text{---}OH \ + \ (CH_3COO)_2O \longrightarrow R\text{---}O\text{-}COCH_3 + \ CH_3CHOOH$$

(B) Alcoholic –OH:

It is confirmed by oxidation reaction.

❖ **Primary –OH:**

Primary alcohol on oxidation gives first an aldehyde and then acid.

$$\text{---}CH_2OH \longrightarrow \text{---}CHO \longrightarrow \text{---}COOH$$

❖ **Secondary –OH:**

Firstly, it forms acid having one carbon less than parent molecule via ketone formation.

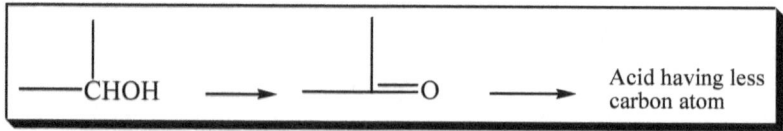

❖ **Tertiary –OH:**

In this method, ter. Alcohol is heated with Grignard reagent and methane is formed which is measured.

$$\text{\Large$\rangle$—OH} \xrightarrow{\text{CH}_3\text{-Mg-I}} \text{\Large$\rangle$—OMgI} + \text{CH}_4$$

(C) Carboxyl (-COOH) Group:

It is identified by,

- Reaction with weak base,
- Formation of ester with alcohol,
- Acid-base titration.

(D) Alkoxy (–OR) group:

Generally alkaloids contain Methoxy (-OCH$_3$) and Ethoxy (-OC$_2$H$_5$) group, which is identified by Zeisel method.

In this method, Alkoxy group is converted to alkyl halide by using HI which is estimated as silver iodide by treatment with ethanolic silver nitrate.

$$\text{—OCH}_3 \xrightarrow[\text{126 °C}]{\text{HI}} \text{—OH} + \text{CH}_3\text{I} \xrightarrow{\text{Alc. AgNO}_3} \text{AgI}$$

(E) Ester (-COOR) & Amide (-CONH$_2$) group:

Ester and related amide, lactone and lactam groups are identify by their hydrolysis with water, acid or alkali, gives hydroxyl group and acidic compounds. A type of product gives information about presence of above group.

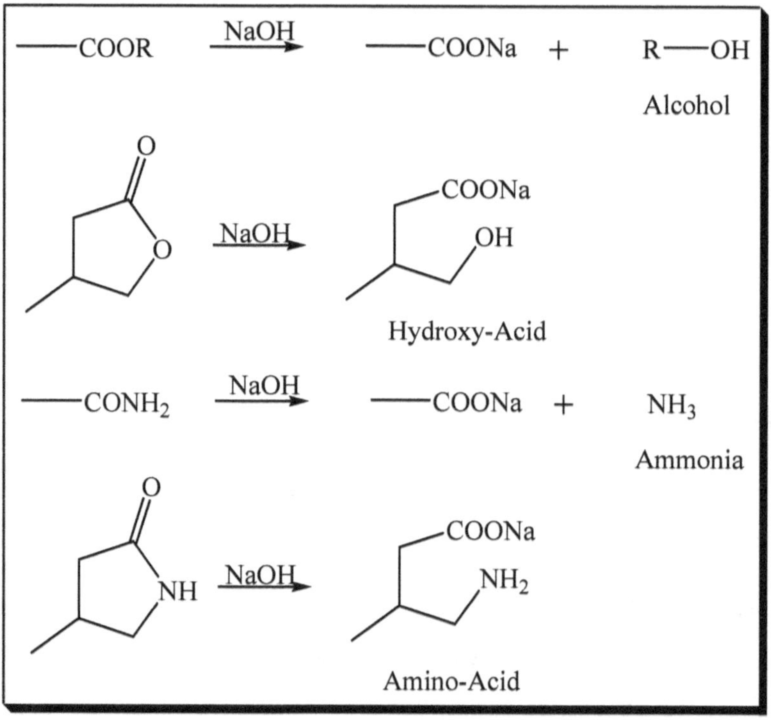

(F) Carbonyl (-CO-) group:

It can be identified by,

Addition product of NaHSO$_3$,

Oximes with hydroxylamine,

Phenyl hydrazone with phenyl hydrazine.

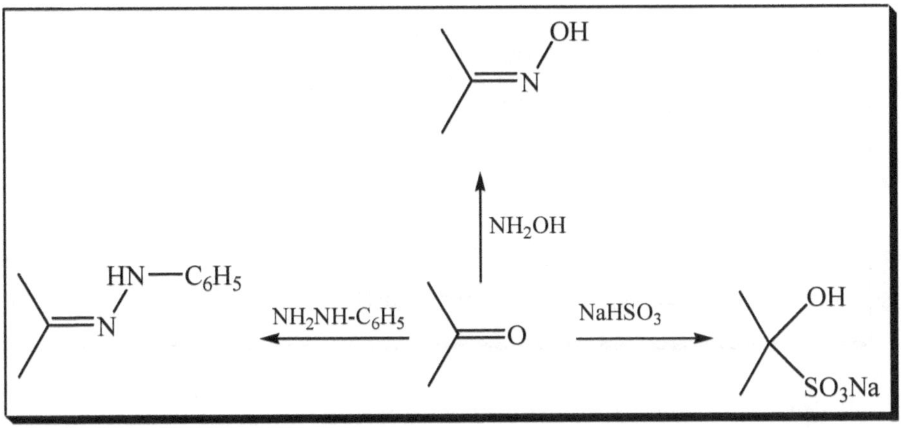

OH

$\diagup$=N$\diagup$ (N—OH)

$\uparrow$ NH$_2$OH

HN—C$_6$H$_5$

$\diagup$=N$\diagup$ ←NH$_2$NH-C$_6$H$_5$— $\diagup$=O —NaHSO$_3$→ $\diagup\!\!\diagup$ OH / SO$_3$Na

Step V: Nature of Nitrogen atom

In majority of alkaloid, the nitrogen atom involve in a ring structure, therefore it must be secondary or tertiary. To differentiate following method are used:

(A) Only secondary amines are acetylated, benzoylated and give Liebermann's nitroso reaction.

The consumption of alkalyl halide for the quaternary salt formation of alkaloids gives good idea about types of amine such as **tertiary amine** require only one mole while secondary amine require two moles of alkyl halide.

$$\underset{C_4H_8}{\overset{C_4H_8}{\diagdown}}NH \quad \xrightarrow[-HI]{2CH_3I} \quad \underset{C_4H_8}{\overset{C_4H_8}{\diagdown}}\overset{\oplus}{N}(CH_3)_2 \ \overset{\ominus}{I}$$

$$N\!\!\equiv\!\!C_{10}H_{14}\!\!\equiv\!\!N \quad \xrightarrow[-HI]{2CH_3I} \quad \overset{\ominus}{HI}(H_3C)_2\overset{\oplus}{N}\!\!\equiv\!\!C_{10}H_{14}\!\!\equiv\!\!\overset{\oplus}{N}(CH_3)_2 \ \overset{\ominus}{I}$$

(B) Ter. Amine gives Nitro oxide on treatment with H_2O_2.

$$-\overset{\diagup}{\underset{\diagdown}{N}} \quad \xrightarrow{H_2O_2} \quad -\overset{\diagup}{\underset{\diagdown}{N}}\!\!\longrightarrow O$$

Step VI: Presence of N-Alkyl Group

(A) An alkaloid on distillation with alkali or soda-lime generally yields, methyl amine, ethyl amine, dimethyl amine, diethyl amine indicates presence of respectively alkyl groups attached with the nitrogen atom.

⁕ Nicotine on heating with alkali, yields one mole methyl amine, indicates presence of one N-Methyl group.

$$(C_9H_{11}N)N\text{-}CH_3 \quad \xrightarrow{\text{Aq. KOH}} \quad CH_3\text{-}NH_2$$

Nicotine   Methyl amine

(B) N-alkyl groups are estimated by Herzig and Meyer method, in which the alkaloids heated with hydroiodic acid at 200-300 °C. In these process alkyl group is convert into the alkyl iodide, which are estimated as silver iodide by means of silver nitrate.

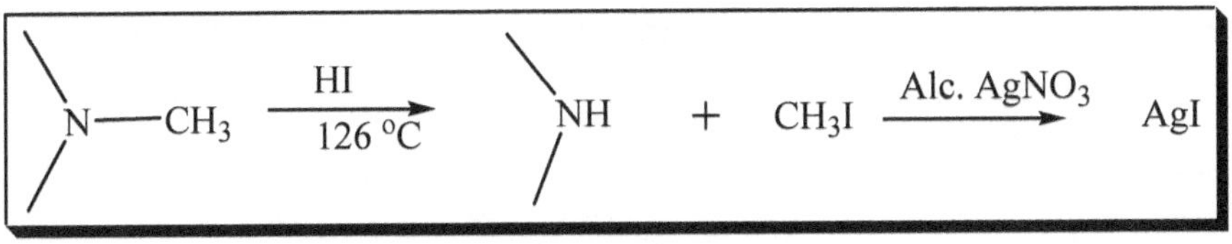

$$\text{N---CH}_3 \xrightarrow[\text{126 °C}]{\text{HI}} \text{NH} + \text{CH}_3\text{I} \xrightarrow{\text{Alc. AgNO}_3} \text{AgI}$$

Step VII: Presence of C- Methyl Group:

C- Methyl groups are estimated by Kuhn-Roth method in which alkaloids are oxidized by chromic acid, it furnished acetic acid, which can be determined by titrimetric method against standard base solution.

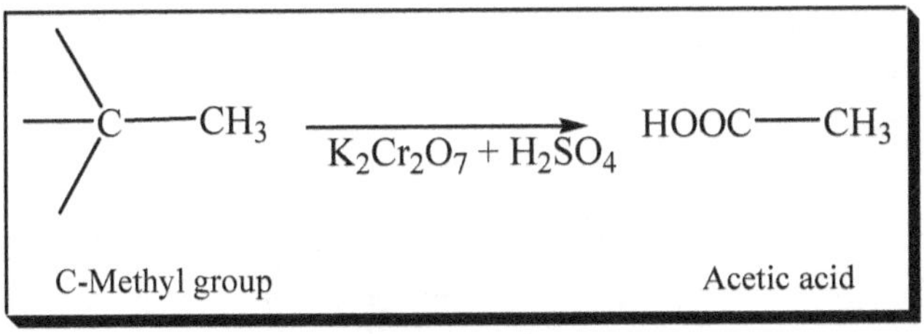

$$\text{C---CH}_3 \xrightarrow{\text{K}_2\text{Cr}_2\text{O}_7 + \text{H}_2\text{SO}_4} \text{HOOC---CH}_3$$

C-Methyl group   Acetic acid

Step VIII: degradation of alkaloids:

It is the most important step to prove the structure of alkaloids. Various degradation methods discussed below:

(A) Emde's Method:

Compounds which are difficult to degrade by other process, this method is used, which consists reduction of quaternary ammonium salt with sodium amalgam or sodium in ammonia.

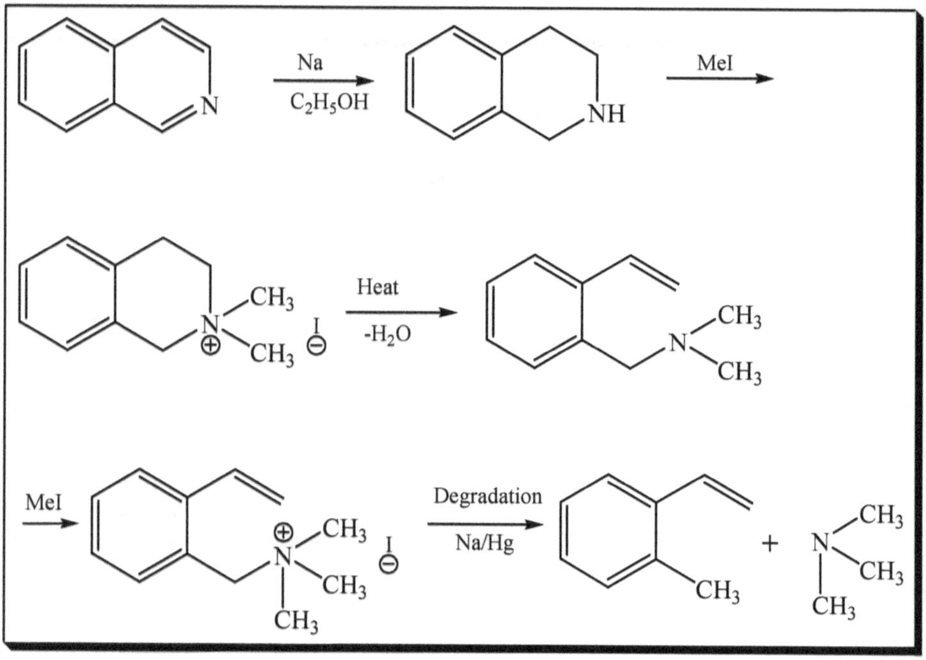

(B) Von Braun Method:

These methods have two categories.

❖ In first category, alkaloids containing nitrogen atom in the ring is treated with cyanogen bromide and later on hydrolysis followed by decarboxylation form brominated sec. amine.

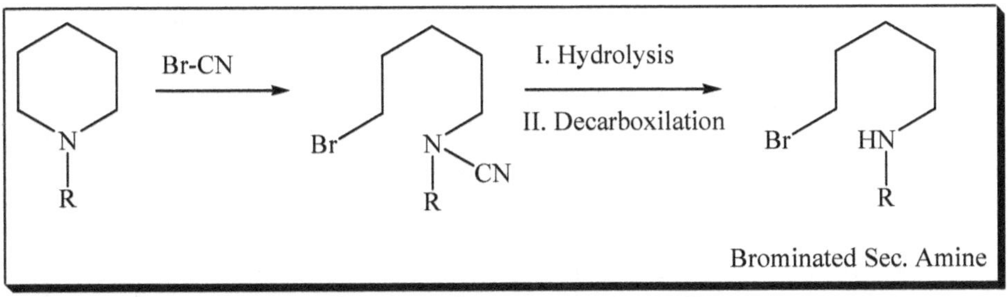

Brominated Sec. Amine

❖ In the second category, secondary cyclic amine reacts with benzoyl chloride, which on reaction with PBr_5 and distillation, nitrogen atom eliminated as benzonitrile with the formation of dihalo- compounds.

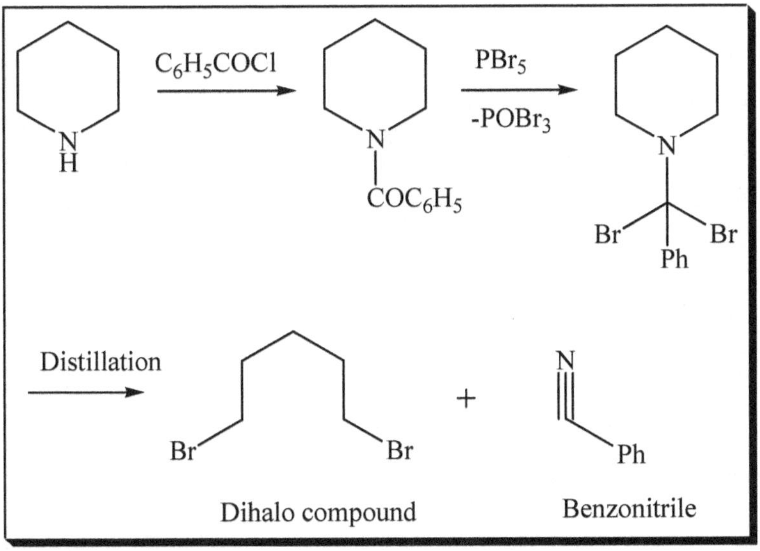

Dihalo compound   Benzonitrile

(C) Reductive degradation:

Heterocyclic ring of alkaloids open when it treated with HI at 300 °C temprature and liberated ammonia.

16

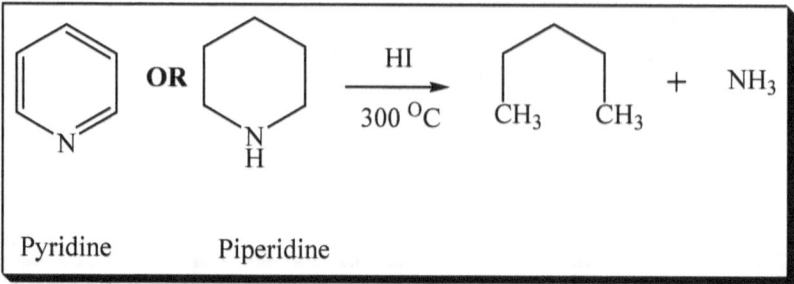

Pyridine   Piperidine

(D) Zinc dust distillation:

Distillation of alkaloids or its product over hot Zn dust, it degrades to stable aromatic compounds. E.g.

- Morphine gives Phenanthrene,
- Connine gives 2-propyl pyridine,
- Cinchonine gives quinoline,
- Papavarine gives isoquinoline, all these indicates presence of respective alkaloids.

(E) Oxidation:

Oxidation by $KMnO_4$, Chromic acid, Nitric acid and H_2O_2, gives useful indication about the structure, the position of nitrogen and oxygen atoms or functional groups like >C=C<, -CHOH, >NH, -OCH_3 etc.

(F) Dehydrogenation:

In this method, alkaloids are distilled with a catalyst like S, Se, and Pd to form relatively simple products which provide idea about the skeleton of the alkaloids.

Step IX: Synthesis:

Finally the structure proposed by degradative methods is confirmed by the synthesis of definite alkaloids.

Step X: Physical methods:

Now a days, the structure of alkaloids has been identified by the use of modern instrumental techniques such as UV, IR, NMR, Mass spectroscopy, X-Ray Diffraction methos, Gas chromatography and HPLC etc.

10.0   PROPERTIES AND SYNTHESIS OF CONINE

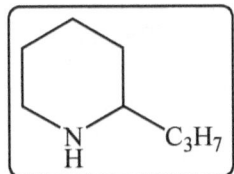

Connine is a Pyridine-Piperidine group alkaloid. It is major constitution of Hemlock alkaloids.

Connine has on their special importance due to their special characteristics such as poisonous properties of hemlock. Presence of Connine in Hemlock is responsible for disturbance of CNS and cause the death by the stop of breathing.

10.1 Properties:

- Poisonous colorless liquid, turns brown on exposure to air,
- Unpleasant odour and strongly basic in nature,
- Boiling point 167 °C,
- Optical activity: Dextro rotatory (Rotation: +15.7 °),
- Soluble in water and organic solvent

10.2 Synthesis:

[1] **Ledenberg synthesis:**

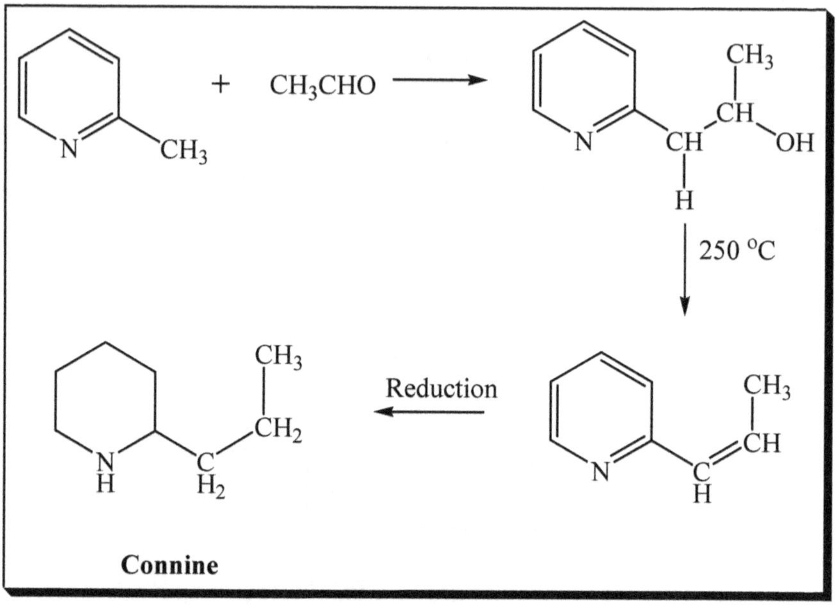

Connine

Bergmann Synthesis:

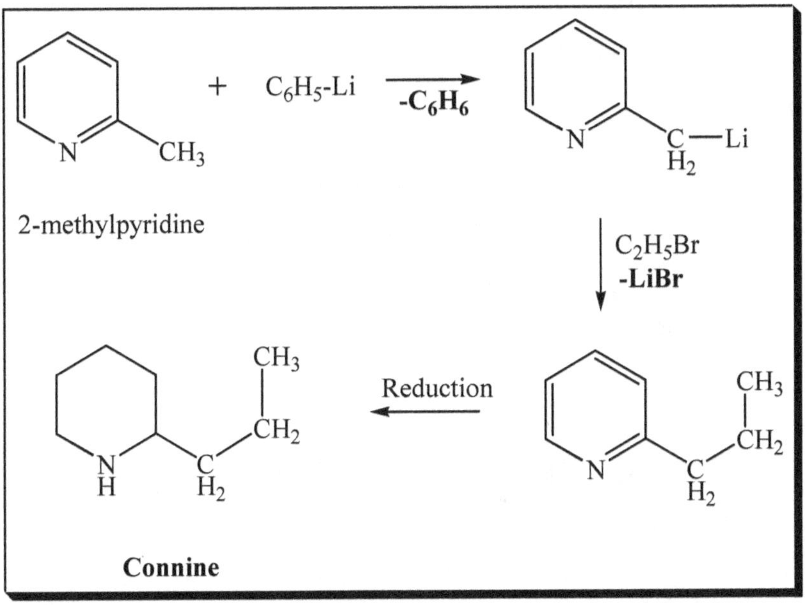

11.0 PROPERTIES AND SYNTHESIS OF NICOTINE

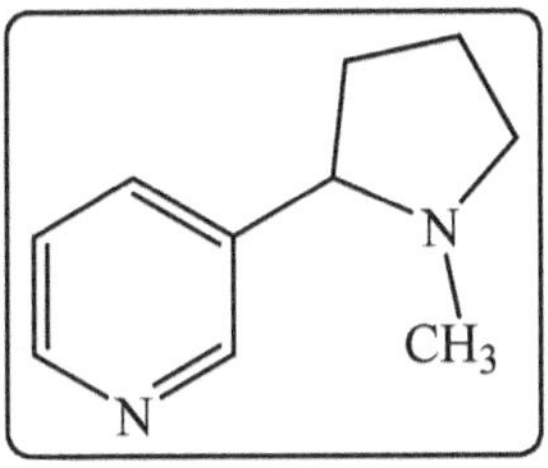

Nicotine is a most important alkaloid found in tobacco. It is pyridine-Pyrolidine alkaloids and most toxic alkaloid, a fatal dose for man being about 40 mg.

Dry tobacco leaves contain about 4 % nicotine combined with citric and malic acid. The free nicotine is obtained by steam distillation and then purified through oxalate.

11.1 Properties:

- Nicotine is colorless liquid; it becomes darkened on exposure to air,
- It is a leavo-rotatory (Optical Rotation: - 166.4 °), but its salt is Dextro-rotatory,
- Boiling point is 246 °C,
- It is very deadly poison,
- It greatly effect Respiratory system,
- Also used in large extent as an insecticide and preparation of nicotinic acid.

11.2 Synthesis:

Spath synthesis:

12.0   PROPERTIES AND SYNTHESIS OF PAPAVARINE

Papavarine is an isoquinoline alkaloid found in opium poppy together with nearly 24 other alkaloids, some of these are Narcotine, morphine, codeine etc.

It was first time isolated from opium in 1848 and occurs as small quantities (0.8 to 1 %). Opium is dried latex obtained from the green buds of the poppy.

12.1 Properties:

- It is colorless liquid,
- Melting point is 147 °C,
- Insoluble in water but soluble in alcohol and chloroform,
- It has more affinity to slow down heart than morphine,
- It is used as an anti-spasmodic and relaxation of cardiac muscles.

12.2 Synthesis:

Synthesis of Papavarine is divided in to two steps;

➢ Synthesis of Homo-veratryl-amine and Homo-veratroyl-chloride, and

➢ Condensation of Homo-veratryl-amine and Homo-veratroyl-chloride.

1. Synthesis of Homo-veratryl-amine and Homo-veratroyl-chloride:

Homo-Veratryl-Amine

Homo-Veratroyl-Chloride

2. Condensation of two compounds I and II:

Papavarine

13.0   CONSTITUTION, PROPERTIES AND SYNTHESIS OF RICININE

It is found to be present in castor-oil-seeds. It is less toxic than other alkaloids. It is very weak base. It is bitter in tastes.

13.1 Constitution:

1   Its molecular formula is $C_8H_8N_2O_2$.

2   On zinc dust distillation, ricinine gives pyridine, indicating that the alkaloids have a pyridine nucleus.

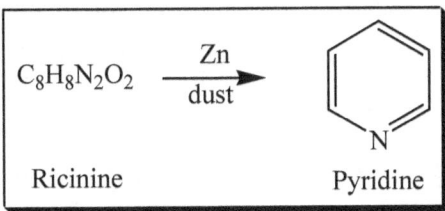

Furthermore, ricinine absorbs two molecules of H_2 on catalytic presence of a dihydrogenated pyridine nucleus.

3   Alkaline hydrolysis of ricinine gives $C_7H_6N_2O_2$, ricinine acid along with the formation of methanol indicating the presence of a methoxy group in ricinine.

$$C_7H_5ON_2OCH_3 \xrightarrow[\text{NaOH}]{\text{Hydrolysis}} C_7H_5ON_2.OH + CH_3\text{-}OH$$

Ricinine   Ricinine acid

Ricinine acid gives red color with $FeCl_2$ and on treatment with phosphorus oxychloride followed by reduction, it gives ricinidine $C_7H_6ON_2$; the changes seem to be happened in the –OH group.

$$C_7H_5ON_2OH \xrightarrow{\text{PCl}_5} C_7H_5ON_2Cl \xrightarrow{\text{Reduction}} C_7H_6ON_2$$

Ricinine acid   Ricinidine

4 The structure of ricinine is based on the structure of ricinidine. Ricinidine may be hydrolyzed in two steps.

In the first step, it is found to give an amide of the composition $C_7H_8O_2N$ which on further hydrolysis (second step) gives on acid $C_7H_7NO_3$ (1) with one mole of ammonia.

The results indicate the presence of a nitrile group. The acid $C_7H_7NO_3$ is found to be 1-methyl 2-pyridon 3-carboxylic acid by synthesis and hence the structure of ricinidine must be (2).

Ricinidine

(1)   (2)

Thus ricinine acid will be (3) respectively.

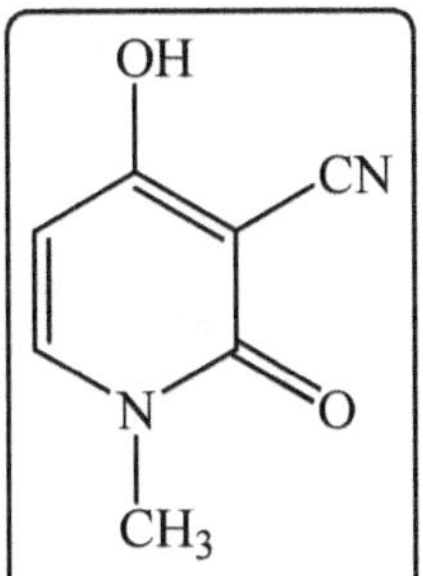

OH

CN

O

N

CH₃

Ricinine acid

(3)

Thus the structure of ricinine is below. Further it is proved by synthesis of ricinine.

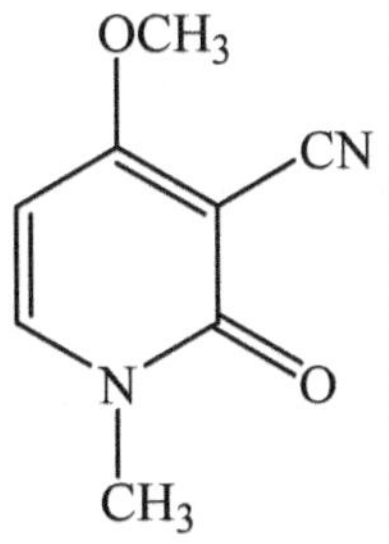

OCH₃

CN

O

N

CH₃

Ricinine

13.2 Synthesis of ricinine:

4-Chloro quinoline — Oxidation KMnO$_4$ → **4-Chloro-pyridine-2,3-dicarboxylic acid** — (CH$_3$CO)$_2$O, -H$_2$O →

NH$_3$ → ... — Hofman Reaction, Br$_2$ + KOH → ... — NaNH$_2$ + H$_2$SO$_4$ →

POCl$_3$, -OH → ... — NH$_3$, -HCl → ...

POCl$_3$, 100 °C, -H$_2$O → ... — 1. CH$_3$OH, CH$_3$ONa 2. Methylation, -HCl, NaCl → ...

$\triangle$, CH$_3$I → **Ricinine**

14.0 CONSTITUTION, PROPERTIES AND SYNTHESIS OF ATROPINE

It is present in the plant ATROPA BELLADONA.

Uses: Atropine is used in ophthaquinology, for this purpose one part of atropine in 13, 000, 0 parts of water is sufficient to cause the dilation of the pupils of eyes of cats.

14.1 Constitution:

1 Its molecular composition is $C_{17}H_{23}NO_3$.

2 On hydrolysis, it gives an alcohol tropine $C_8H_{10}NO$ (±) and tropic acid $C_9H_{10}O_3$ indicating that atropine is an ester.

$$C_{17}H_{12}NO_3 + H_2O \longrightarrow C_8H_{15}NO + C_9H_{10}O_3$$

Atropine   Tropine   Tropic acid

3 If, we know the constitution of tropic acid & tropine than it is easy to find out the structure of atropine.

14.2 Constitution of tropic acid:

1 It's molecular formula is $C_9H_{10}O_3$.

2 Tropic acid does not give addition product with Br_2. It is saturated compound.

3 Silver chloride test says that, it is mono carboxylic acid derivatives.

4 Tropic acid on benzylation and acetylation gives monobenzyl and acetyl derivatives respectively.

$$R\text{-OH} + C_6H_5COCl \xrightarrow[-HCl]{} C_6H_5COOR$$

$$R\text{-OH} + CH_3COCl \xrightarrow[-HCl]{} CH_3COOR$$

It proves that -OH group is present in the structure of tropic acid.

5 On strong heating, it loses a molecule of water to form atropic acid, $C_9H_8O_2$, which on oxidation gives benzoic acid.

$$C_9H_{10}O_3 \xrightarrow[-H_2O]{} C_9H_8O_2 \xrightarrow{[O]}$$

Tropic acid   Atropic acid   Benzoic acid

The formation of benzoic acid suggests that atropic acid and hence tropic acid both contain a benzene ring with a one side chain.   '

Therefore, the major four possible structure of tropic acid are obtained.

Structure (1) and (2) on dehydrogenation will give cinamic acid.

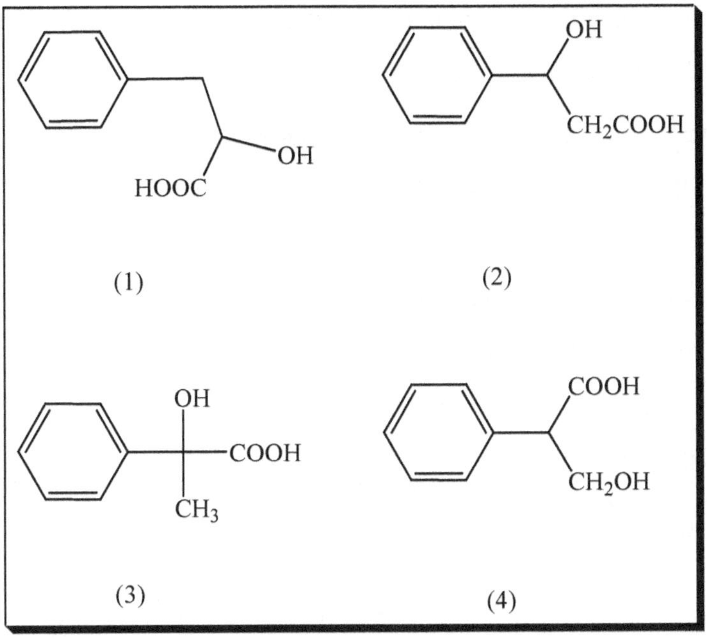
OH
HOOC
(1)
OH
CH2COOH
(2)
OH
COOH
CH3
(3)
COOH
CH2OH
(4)

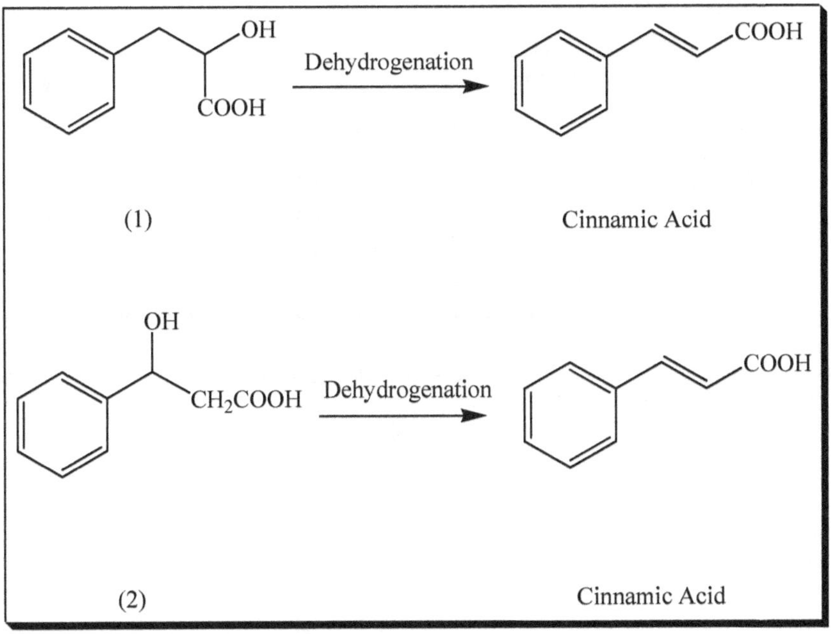
OH
COOH
Dehydrogenation
COOH
(1)
Cinnamic Acid
OH
CH2COOH
Dehydrogenation
COOH
(2)
Cinnamic Acid

Practically cinnamic acid is not obtain, so structure 1 and 2 are not the true structure of tropic acid.

Structure (3) was synthesized by the scientist Mackenzie and wood from acetophenone but its melting point, and others physical properties are not identical with the tropic acid.

So, Structure (3) is also not a structure of tropic acid. It was identifying as an atrolactic acid, which is nothing but the structure of numbered (3).

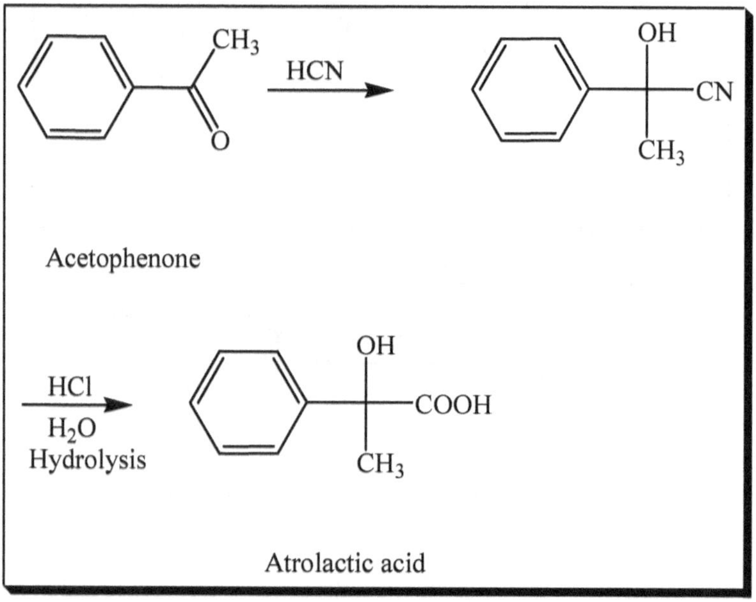

Acetophenone

Atrolactic acid

Then, atrolactic acid was further converted into tropic acid.

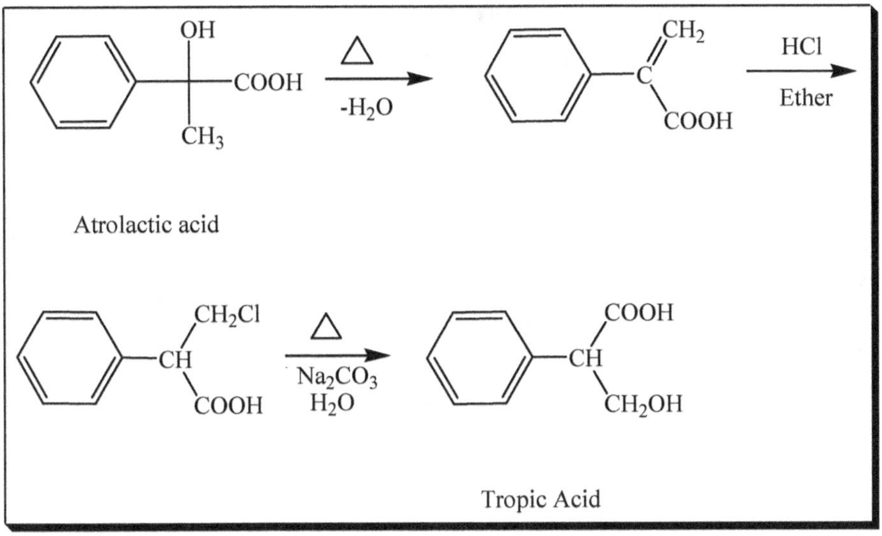

Atrolactic acid

Tropic Acid

14.3 Synthesis of tropic acid:

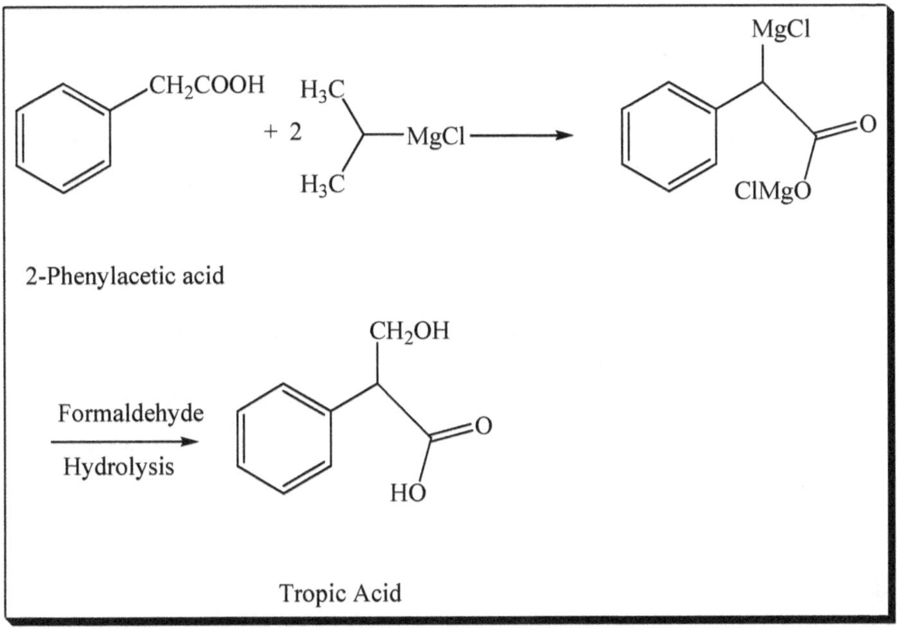

2-Phenylacetic acid

Tropic Acid

14.4 Constitution of tropine (Tropanol):

1 Its molecular formula is $C_8H_{15}NO$.

2 It is saturated secondary alcohol having tertiary nitrogen in the form of $-NCH_3$ group.

3 BENZYLATION &ACETYLATION:

Benzylation & acetylation gives mono benzyl &mono acetyl derivatives. So, -OH group is present.

4

$$C_8H_{15}NO \xrightarrow[-H_2O]{\text{Con. } H_2SO_4} C_8H_{13}N$$

Tropine   Tropidine

So, Secondary –OH group is present.

5

$$C_8H_{15}NO \xrightarrow[CrO_3]{[O]} C_8H_{13}NO \xrightarrow{2\ C_8H_5CHO} \text{Di-Benzylidine Derivati}$$

Tropine   Tropinone

Formation of ketone indicate that –OH group is secondary [>CH-OH]. Formation of dibenzylidene derivative confirms that methylene group is present. So, $-CH_2-CO-CH_2-$ group present in tropinone and $-CH_2-CHOH-CH_2-$ group is present in tropine.

6

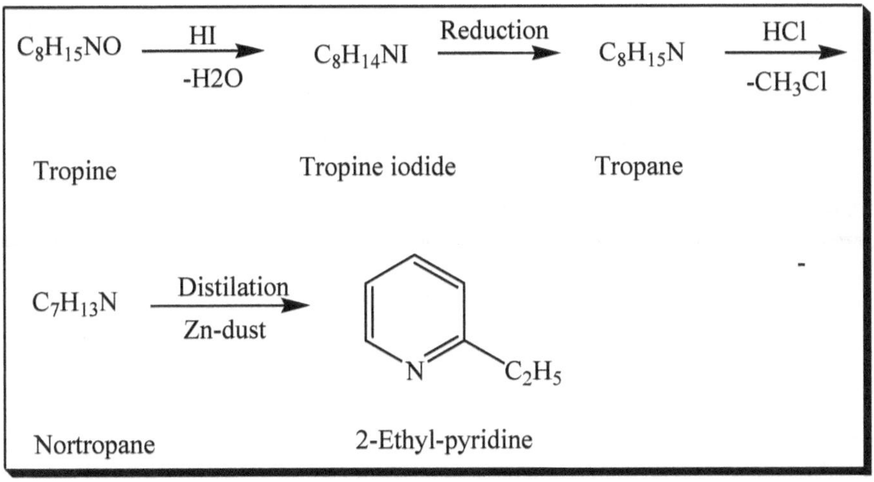

$$C_8H_{15}NO \xrightarrow[-H2O]{HI} C_8H_{14}NI \xrightarrow{Reduction} C_8H_{15}N \xrightarrow[-CH_3Cl]{HCl}$$

Tropine   Tropine iodide   Tropane

$$C_7H_{13}N \xrightarrow[Zn\text{-}dust]{Distilation}$$

2-Ethyl-pyridine

Nortropane   2-Ethyl-pyridine

So, the formation of 2-ethylpyridine indicate that reduced pyridine ring is present in tropine.

7

$$C_8H_{15}NO \xrightarrow[CrO_3]{[O]} C_8H_{13}NO \xrightarrow[CrO_3]{[O]} C_8H_{13}NO_4$$

Tropine   Tropinone   Tropinic acid

Tropinic acid is obtained from ketone and, there is no loss of carbon atom ($C_8 \rightarrow C_8$). So, -OH group is present in ring system, and ketone must be cyclic ketone.

8

$C_8H_{13}NO_4$ $\xrightarrow[\substack{\text{Methylation} \\ \text{(C-N linking react)}}]{\text{Halfmann -exostick}}$ $C_7H_8O_4$ $\xrightarrow[4H]{\text{Reduction}}$ $HOOC\text{-}(CH_2)_5\text{-}COOH$

Tropinic acid   Piperylene carboxylic acid   Pimelic acid

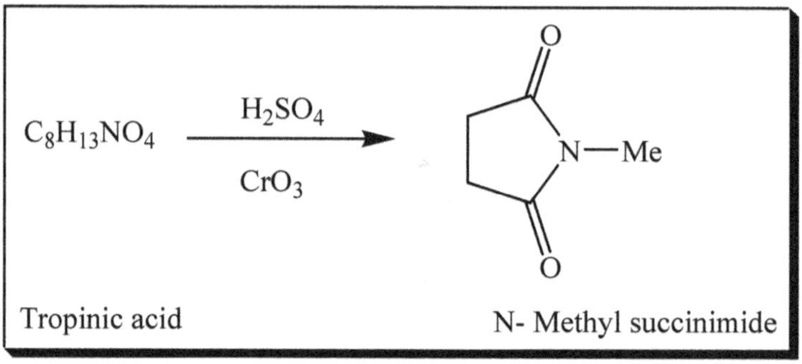

$C_8H_{13}NO_4$ $\xrightarrow[CrO_3]{H_2SO_4}$

Tropinic acid   N- Methyl succinimide

Formation of N-methyl succinimide indicates that 5-member ring should be present. Formation of pimelic acid indicates that 7-member ring present should be present. And so the structure of tropine must be (1).

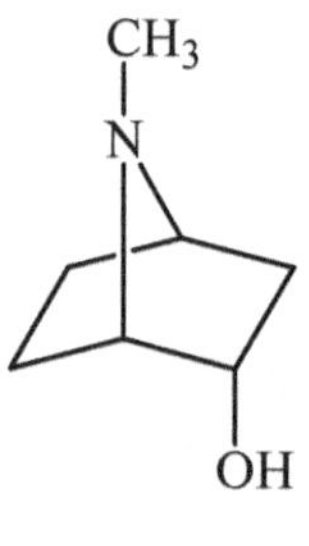

(1)

(1)

The above proposed structure for tropine is confirmed by its degradation reaction.

As we know that,

Acid + Tropine ⟶ Atropine

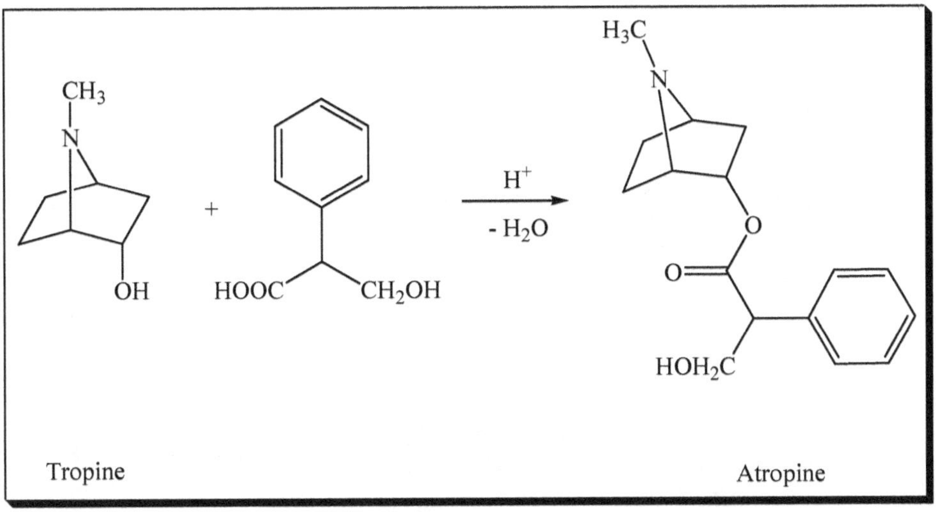

Tropine

Atropine

15.0 SYNTHESIS OF NARCOTINE

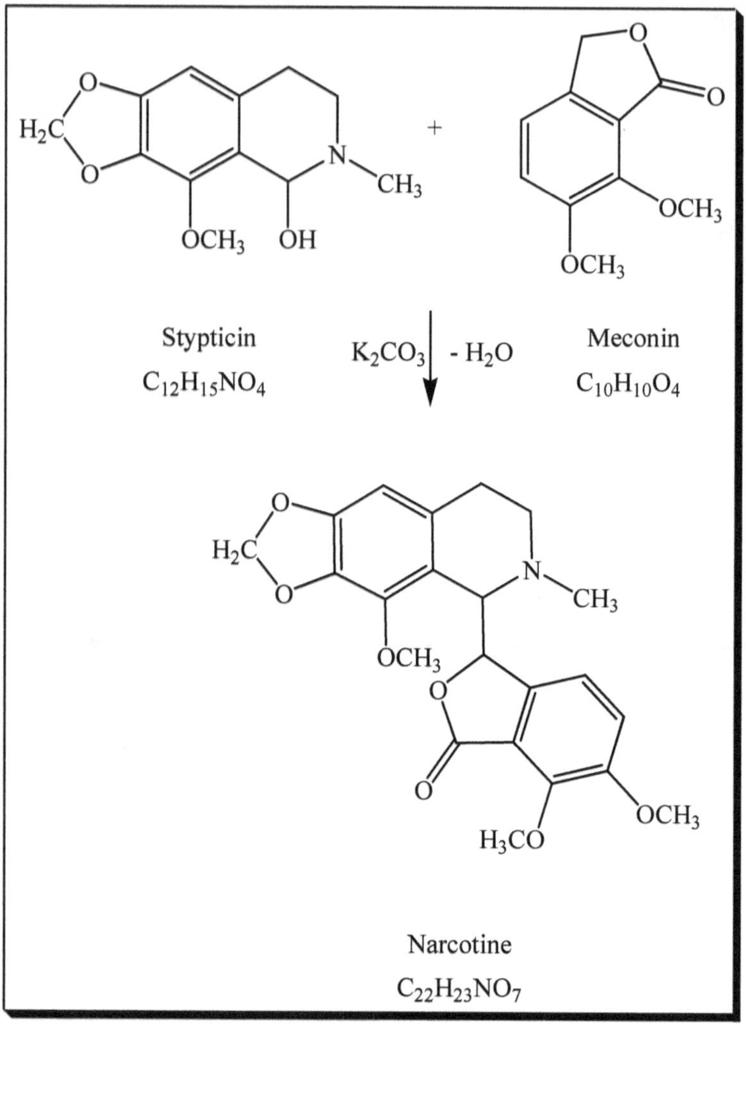

Stypticin

$C_{12}H_{15}NO_4$

K_2CO_3 $- H_2O$

Meconin

$C_{10}H_{10}O_4$

Narcotine

$C_{22}H_{23}NO_7$

16.0   SYNTHESIS OF MORPHINE:

Naphthalene-2,6-diol

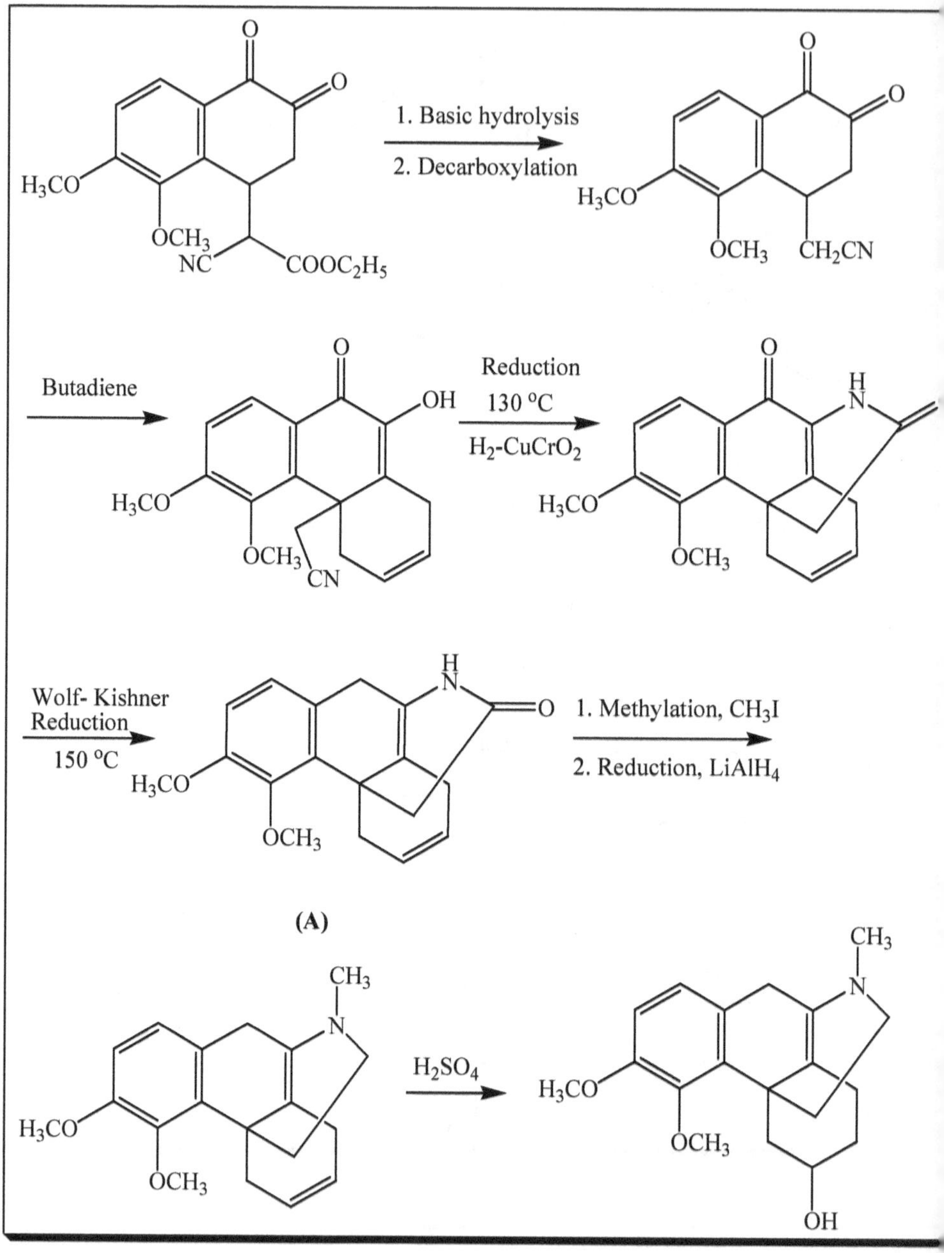
H3CO
OCH3
NC
COOC2H5
1. Basic hydrolysis
2. Decarboxylation
CH2CN
Butadiene
OH
CN
Reduction
130 °C
H2-CuCrO2
Wolf- Kishner
Reduction
150 °C
1. Methylation, CH3I
2. Reduction, LiAlH4
(A)
CH3
H2SO4
OH

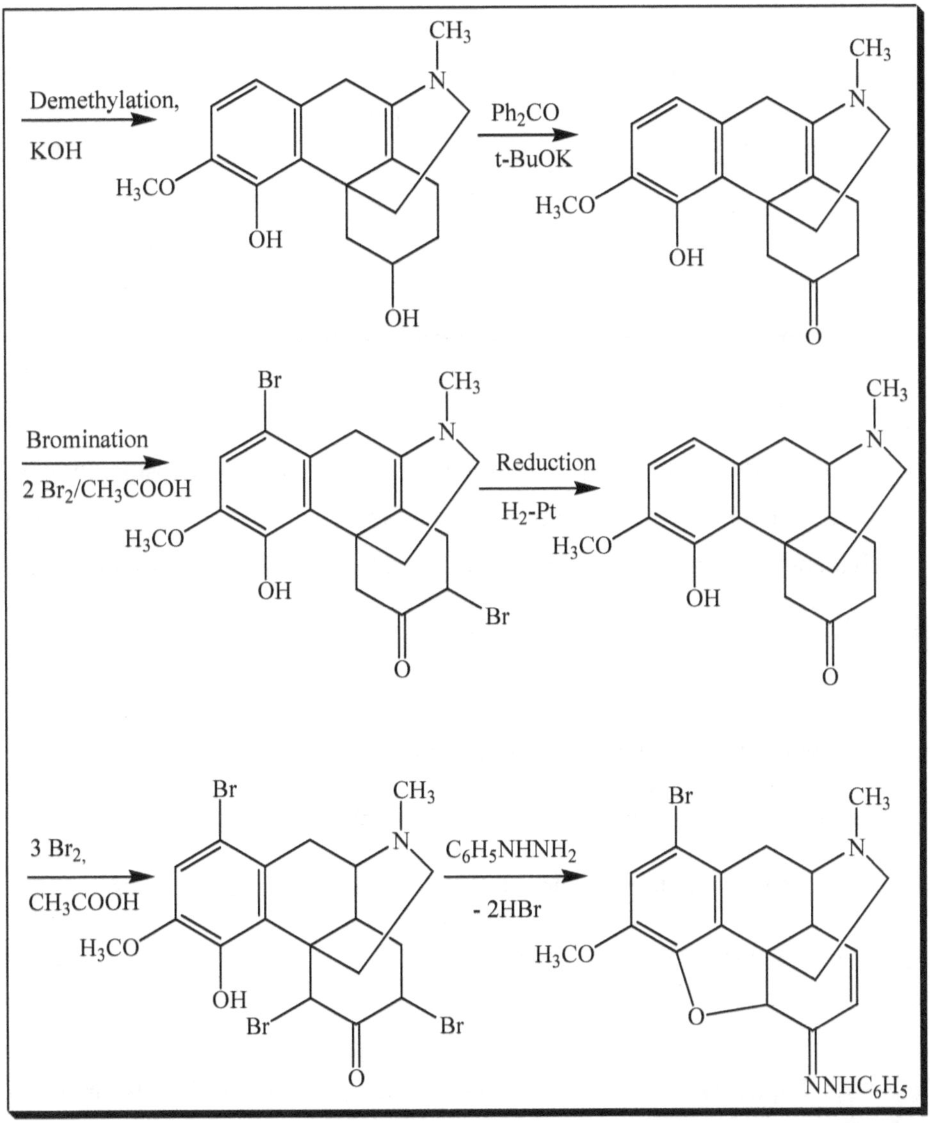

Demethylation,
KOH
CH3
N
H3CO
OH
OH
Ph2CO
t-BuOK
CH3
N
H3CO
OH
O
Bromination
2 Br2/CH3COOH
Br
CH3
N
H3CO
OH
Br
O
Reduction
H2-Pt
CH3
N
H3CO
OH
O
3 Br2,
CH3COOH
Br
CH3
N
H3CO
OH
Br
Br
O
C6H5NHNH2
- 2HBr
Br
CH3
N
H3CO
O
NNHC6H5

1. Hydrolysis,
-$C_6H_5NHNH_2$
- HBr

2. Reduction

CH_3
N
H_3CO
O
OH

C_5H_5N/HCl
220 °C

CH_3
N
HO
O
OH

17.0   OPIUM POPPY

Oldest evidence of poppy uses was existence of a poppy capsule, founded with religious artifacts from in spain, that are dated 7500 years old. Evidence of spread was mainly founded in Europe from 3000 to 5000 years ago.

17.1 Opium Poppy (Papaver somniferum) Properties:

Member of the papaveraceae, poppy family,

Large showy annual with flowers (white, pink, red, purple),

Fruit is a capsule that produces latex with several potent alkaloids,

Latex is collected from cut capsules.

17.2 Ancient medical use:

− Poppy was revered by several ancient societies for its analgesic properties and its ability to induce sleep,
− Opium latex has a long history of use for pain relieve and inducing sleep among Babylonia, Egyptian, Greek, and Roman civilizations,
− Opium eaten, drunk, and smoked,
− Most common method was to dissolve opium in alcohol – opium wine

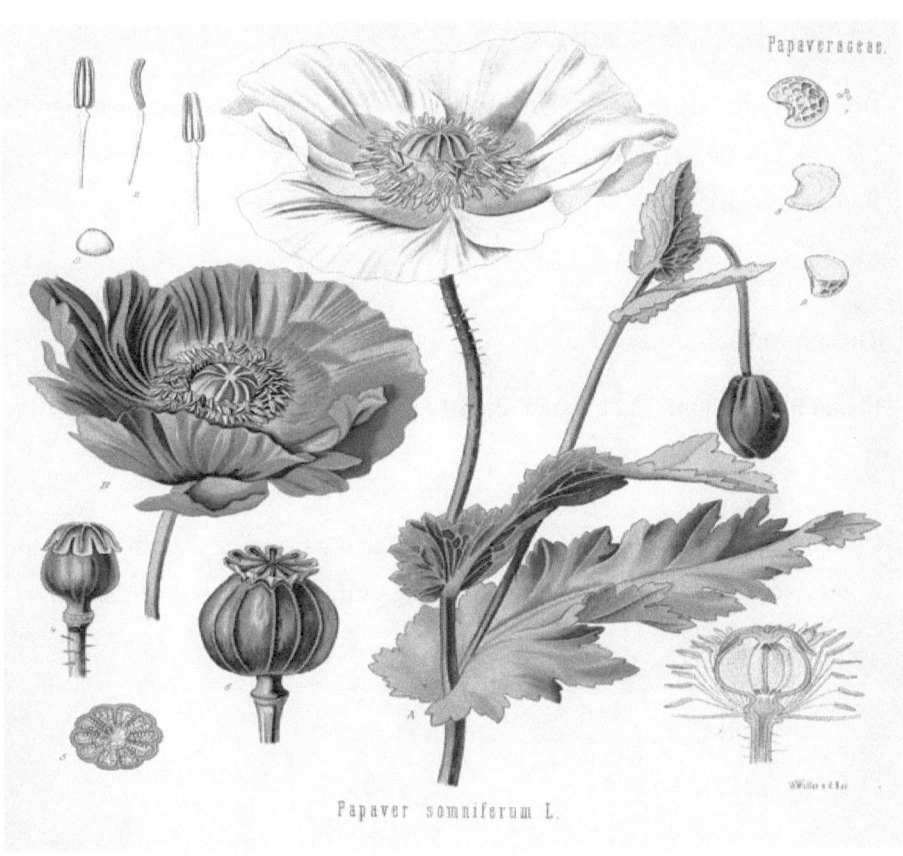

Papaveraceae.

Papaver somniferum L.

17.3 OPIUM ALKALOIDS

17.3.1 Occurrence:

Opium is the air-dried milky exudate, or latex, obtained by incising the mature unripe capsules of the opium poppy *Papaver somniferum* (Fam. Papaveraceae).

17.3.2 Sources:

Turkey, India, Europe.

Illegal Production: Afghanistan, Pakistan, Iran, Burma, Laos, Thailand.

17.3.3 Constituent of Opium:

Opium contain more than 40 alkaloids usually combined with a specific acid (Meconic acid) or with other acids e.g. sulfuric and acetic acids.

17.3.4 Six main opium alkaloids are:

- Morphine
- Codeine
- Thebaine
- Noscapine
- Narceine
- Papavarine

17.3.5 Classification:

Opium alkaloids can be sub classified into 3 main groups with different basic nuclei:

- Phenanthrene alkaloids e.g. Morphine and Codeine.

- Benzylisoquinoline alkaloids e.g. Papavarine and Noscapine (Narcotine).

- Phenyl ethylamine alkaloids e.g. Nerceine.

18.0   MORPHINE

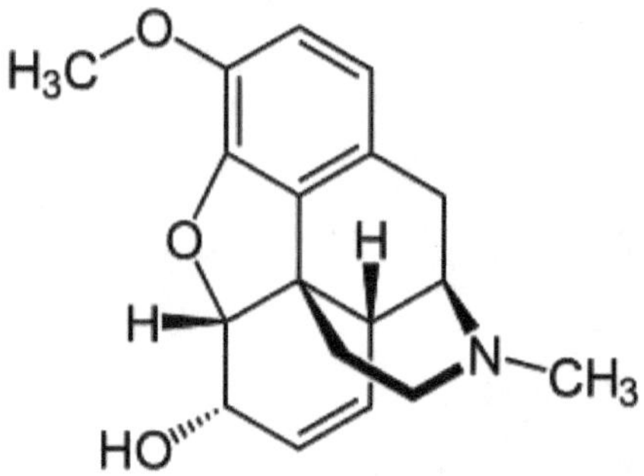

Morphine is soluble in sodium hydroxide due to phenolic group present.

Morphine is levorotatory, insoluble in water, sparingly soluble in ethanol and chloroform, practically insoluble in ether and benzene.

18.1 Uses:

Morphine act as a narcotic analgesic (reduce pain & induce sleep) in a dose of 5-20 mg of morphine hydrochloride, sulfate or tartrate, administered orally or parentally, every 4 hours.

Used before and after surgical operations and to terminally ill cancer patients.

Suppress movement, so stops diarrhea.

18.2 Adverse Effects:

Two major problems are associated to use of morphine: Addiction and Tolerance.

Morphine affects the central nervous system but also induces drowsiness and can depress respiration, overdose can cause death through respiratory failure cause has high physical dependency.

Has relatively low oral activity.

Became drug of choice to treat war injuries during Civil War- created many addicts.

19.0   CODEINE:

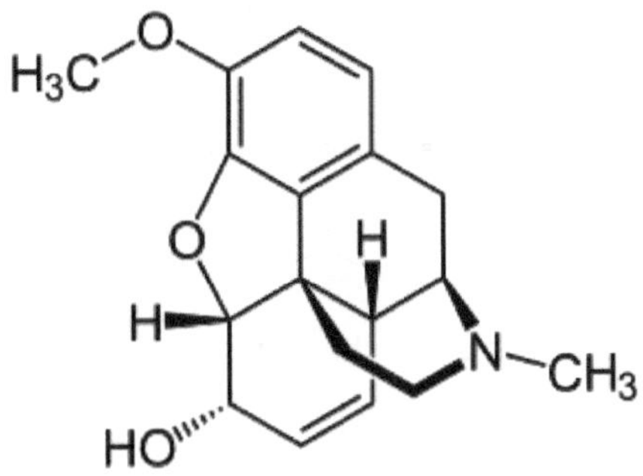

19.1 Properties:

It is soluble in water, boiling water, ethanol, chloroform and ether. Codeine is non-phenolic compound.

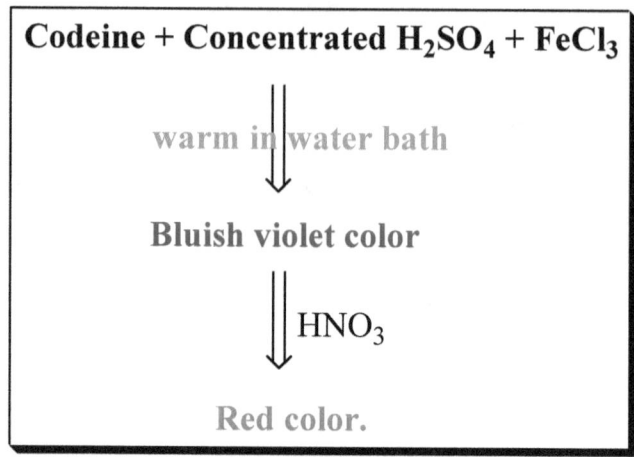

- It has less narcotic analgesic than morphine.

- It is mainly used as antitussive.

- Codeine most widely used opiate.

- Only 1/5 as strong as morphine and still addictive.

- Effective in oral medication and often used in combination with non-opiate compounds.

- Codeine is especially effective in cough syrups because it suppresses the coughing reflex.

20.0   HEROIN:

In 1898, Bayer Pharmaceuticals introduced heroin which they thought was a non-addictive opiate and more effective than morphine and codeine. Heroin is a semi-synthetic derivative of morphine; Diacetyl morphine. It is widely available in over the counter tonics and cough syrups from 1898 to 1914.

Within a few years over 1 million people addicted to heroin in US alone. Soon realized that, it was actually more addictive than morphine- actually six times more addictive.

In US, use of heroin is illegal placed under Harrison Act of 1914. Heroin still used medicinally in other countries. Heroin addiction becomes a major problem over worldwide.

20.1 Use:

The original trade name of heroin is typically used in non-medical settings. It is used as a recreational drug for the euphoria it induces. Anthropologist Michael Agar once described heroin as "the perfect whatever drug.

Tolerance develops quickly, and increased doses are needed in order to achieve the same effects. Its popularity with recreational drug users, compared to morphine, reported stems from its perceived different effects. In particular, users report an intense rush, an acute transcendent state of euphoria, which occurs while diamorphine is being metabolized into 6-monoacetylmorphine (6-MAM) and morphine in the brain. Some believe that heroin produces more euphoria than other opioids; one possible explanation is the presence of 6-monoacetylmorphine, a metabolite unique to heroin – although a more likely explanation is the rapidity of onset. While other opioids of recreational use produce only morphine, heroin also leaves 6-MAM, also a psycho-active metabolite. However, this perception is not supported by the results of clinical studies comparing the physiological and subjective effects of injected heroin and morphine in individuals formerly addicted to opioids; these subjects showed no preference for one drug over the other. Equipotent injected doses had comparable action courses, with no difference in subjects' self-rated feelings of euphoria, ambition, nervousness, relaxation, drowsiness, or sleepiness.

Short-term addiction studies by the same researchers demonstrated that tolerance developed at a similar rate to both heroin and morphine. When compared to the opioids hydromorphone, fentanyl, oxycodone, and pethidine (meperidine), former addicts showed a strong preference for heroin and morphine, suggesting that heroin and morphine are particularly susceptible to abuse and addiction. Morphine and heroin were also much more likely to produce euphoria and other positive subjective effects when compared to these other opioids.

Some researchers have attempted to explain use of heroin and the culture that surrounds it through the use of sociological theories. In Righteous Dopefiend, Philippe Bourgois and Jeff Schonberg use anomie theory to explain why people begin using heroin. By analyzing a community in San

Francisco, they demonstrated that use of heroin use was caused in part by internal and external factors such as violent homes and parental neglect. This lack of emotional, social, and financial support causes strain and influences individuals to engage in deviant acts, including heroin usage.

They further found that heroin users practiced "retreatism", a behavior first described by Howard Abadinsky, in which those suffering from such strain reject society's goals and institutionalized means of achieving them.

20.2 Routes of administration:

Oral

Oral use of heroin is less common than other methods of administration, mainly because there is little to no "rush", and the effects are less potent. Heroin is entirely converted to morphine by means of first-pass metabolism, resulting in deacetylation when ingested. Heroin's oral bioavailability is both dose-dependent (as is morphine's) and significantly higher than oral use of morphine itself, reaching up to 64.2% for high doses and 45.6% for low doses; opiate-naive users showed far less absorption of the drug at low doses, having bioavailabilities of only up to 22.9%. The maximum plasma concentration of morphine following oral administration of heroin was around twice as much as that of oral morphine.

Injection

Injection, also known as "slamming", "banging", "shooting up", "digging" or "mainlining", is a popular method which carries relatively greater risks than other methods of administration. Heroin base (commonly found in Europe), when prepared for injection, will only dissolve in water when mixed with an acid (most commonly citric acid powder or lemon juice) and heated. Heroin in the east-coast United States is most commonly found in the hydrochloride salt form, requiring just water (and no heat) to dissolve. Users tend to initially inject in the easily accessible arm veins, but as these veins collapse over time, users resort to more dangerous areas of the body, such as the femoral vein in the groin. Users who have used this route of administration often develop a deep vein thrombosis.[medical citation needed] Intravenous users can use a various single dose range using a hypodermic needle. The dose of heroin used for recreational purposes is dependent on the frequency and level of use: thus a first-time user may use between 5 and 20 mg, while an established addict may require several hundred mg per day. As with the injection of any drug, if a group of users share a common needle without sterilization procedures, blood-borne diseases, such as HIV or hepatitis, can be transmitted. The use of a common dispenser for water for the use in the preparation of the injection, as well as the sharing of spoons and/or filters can also cause the spread of blood-borne diseases. Many countries now supply small sterile spoons and filters for single use in order to prevent the spread of disease.

Smoking

Smoking heroin refers to vaporizing it to inhale the resulting fumes, not burning it to inhale the resulting smoke. It is commonly smoked in glass pipes made from glassblown Pyrex tubes and light bulbs. It can also be smoked off aluminium foil, which is heated underneath by a flame and the resulting smoke is inhaled through a tube of rolled up foil, This method is also known as "chasing the dragon" (whereas smoking methamphetamine is known as "chasing the white dragon").

Insufflation

Another popular route to intake heroin is insufflation (snorting), where a user crushes the heroin into a fine powder and then gently inhales it (sometimes with a straw or a rolled-up banknote, as with cocaine) into the nose, where heroin is absorbed through the soft tissue in the mucous membrane of the sinus cavity and straight into the bloodstream. This method of administration redirects first-pass metabolism, with a quicker onset and higher bioavailability than oral administration, though the duration of action is shortened. This method is sometimes preferred by users who do not want to prepare and administer heroin for injection or smoking, but still experience a fast onset. Snorting heroin becomes an often unwanted route, once a user begins to inject the drug. The user may still get high on the drug from snorting, and experience a nod, but will not get a rush. A "rush" is caused by a large amount of heroin entering the body at once. When the drug is taken in through the nose, the user does not get the rush because the drug is absorbed slowly rather than instantly.

Suppository

Little research has been focused on the suppository (anal insertion) or pessary (vaginal insertion) methods of administration, also known as "plugging". These methods of administration are commonly carried out using an oral syringe. Heroin can be dissolved and withdrawn into an oral syringe which may then be lubricated and inserted into the anus or vagina before the plunger is pushed. The rectum or the vaginal canal is where the majority of the drug would likely be taken up, through the membranes lining their walls.

20.3 Chemistry; Detection in body fluids:

The major metabolites of diamorphine, 6-MAM, morphine, morphine-3-glucuronide and morphine-6-glucuronide, may be quantitated in blood, plasma or urine to monitor for abuse, confirm a diagnosis of poisoning or assist in a medicolegal death investigation.

Most commercial opiate screening tests cross-react appreciably with these metabolites, as well as with other biotransformation products likely to be present following usage of street-grade diamorphine such as 6-acetylcodeine and codeine. However, chromatographic techniques can easily distinguish and measure each of these substances.

When interpreting the results of a test, it is important to consider the diamorphine usage history of the individual, since a chronic user can develop tolerance to doses that would incapacitate an opiate-naive individual, and the chronic user often has high baseline values of these metabolites in his system.

Furthermore, some testing procedures employ a hydrolysis step before quantitation that converts many of the metabolic products to morphine, yielding a result that may be 2 times larger than with a method that examines each product individually.

21.0 STRUCTURAL ELUCIDATION OF RESERPINE:

21.1 Introduction:

Reserpine is the main constituent of Rauwolfia species, particularly R. Serpentina and R. Vomitoria.

It is mainly used for the treatment of hypertension, headache, tension, asthma and dermatological disorders.

21.2 Constitution of reserpine:

1 Molecular formula: $C_{33}H_{40}N_2O_9$.

2 Presence of five methoxy groups:

By Zeisel method, i.e. when treated with HI yields five molecules of methyl iodide indicating the presence of five methoxy groups.

3 Nature of 'N' atom:

a) Secondary 'N': Formation of monoacetyl derivative with acetic anhydride indicates secondary 'N'.

b) Tertiary 'N': Reserpine forms quaternary ammonium salts with CH_3I, which indicating one of 'N' is tertiary.

4 Hydrolysis: Hydrolysis of reserpine yields mixture of,

i) Methyl alcohol

ii) 3, 4, 5-Trimethoxy benzoic acid

iii) Acid (A) of composition

$$C_{33}H_{40}N_2O_9 + 2H_2O \xrightarrow[\text{Hydrolysis}]{\text{NaOH}} CH_3OH + C_{22}H_{28}N_2O_5 +$$

- As reserpine does not contain -COOH and -OH groups, hence its hydrolysis product reveal that reserpine is a diester.

- The ester linkage in reserpine has been further confirmed by its reduction with lithium aluminum hydride.

$$C_{33}H_{40}N_2O_9 \xrightarrow{\text{LiAlH}_4} C_{22}H_{30}N_2O_4 +$$

Reserpic alcohol

5 **Structure of Reserpic acid:**

A Molecular formula: $C_{22}H_{28}N_2O_5$.

B Presence of one carboxyl group: By usual tests e.g. silver salt method.

C Presence one –OH group: Reserpic acid on oxidation yields a ketone that means it has secondary alcoholic group.

D Nature of two methoxy groups: By Zeisel method.

E Nature of two 'N' atom: It shown that, it contains two 'N' atoms in heterocyclic ring in the form of,

i) Secondary 'N',

ii) Tertiary amino group.

F Reduction of reserpic acid: On reduction with $LiAlH_4$ yields reserpic alcohol.

G Oxidation of reserpic acid: On oxidation with $KMnO_4$, it gives 4-methoxy N-oxalylanthranilic acid.

Reserpic acid   4-methoxy N-oxalyl
                                 Anthranilic acid

Thus one methoxy group is present in *meta* position to –NH group.

H Fusion with KOH: When reserpic acid is fused with potash it yields 5-hydroxy isophthalic acid. One of the acidic groups of isophthalic acid must be present in *m*- position to each other this confirm by the fact that reserpic acid when heated with acetic anhydride yields a gamma lactone.

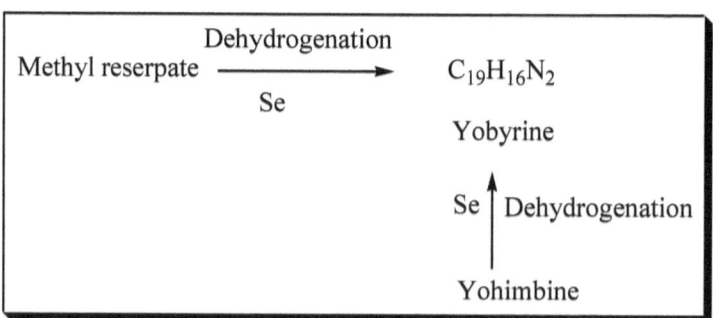

$C_{22}H_{28}N_2O_5$ $\xrightarrow[\text{Fusion}]{\text{KOH}}$

COOH

HO   OH

5-hydroxy isophthalic acid

$C_{22}H_{28}N_2O_5$ $\xrightarrow[\text{AC}_2\text{O}]{\triangle}$

O

Lactone

I Dehydrogenation: When methylreserpate is dehydrogenated with selenium, it yields a hydrocarbon of molecular formula, $C_{19}H_{16}N_2$. This hydrocarbon is also obtained by dehydrogenation of Yohimbine with selenium and, was therefore named as Yobyrine.

Methyl reserpate $\xrightarrow[\text{Se}]{\text{Dehydrogenation}}$ $C_{19}H_{16}N_2$

Yobyrine

Se $\uparrow$ Dehydrogenation

Yohimbine

J Structure of Yobyrine:

i Molecular formula: $C_{19}H_{16}N_2$.

ii Zinc Distillation:

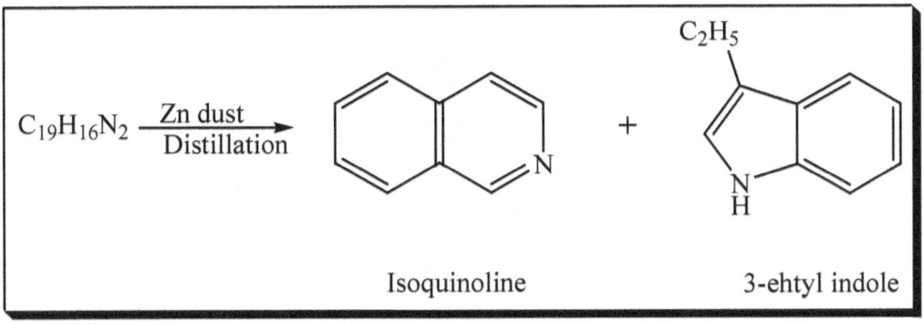

$$C_{19}H_{16}N_2 \xrightarrow[\text{Distillation}]{\text{Zn dust}}$$

Isoquinoline   +   3-ehtyl indole

iii Oxidation: When Yobyrine is oxidized with permanganate, it yields pthalic acid.

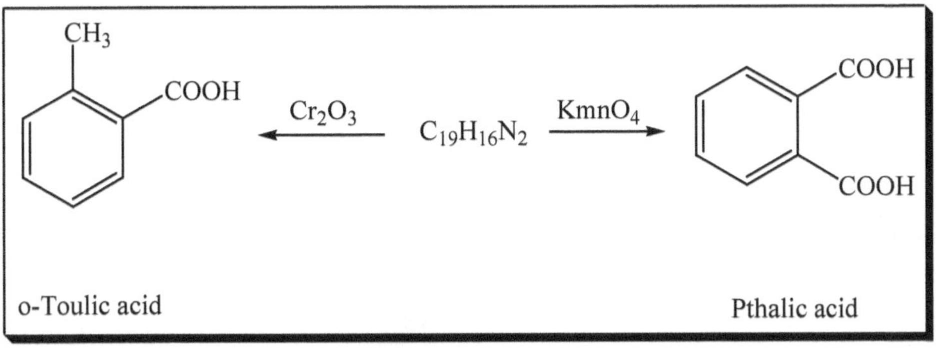

$$\text{o-Toulic acid} \xleftarrow{Cr_2O_3} C_{19}H_{16}N_2 \xrightarrow{KmnO_4} \text{Pthalic acid}$$

o-Toulic acid   Pthalic acid

iv Condensation with aldehydes: Yobyrine gives condensation products, suggesting the presence of pyridine ring with a methylene substitution adjacent to the nitrogen. On the basis of above facts following structure have been postulated for Yobyrine.

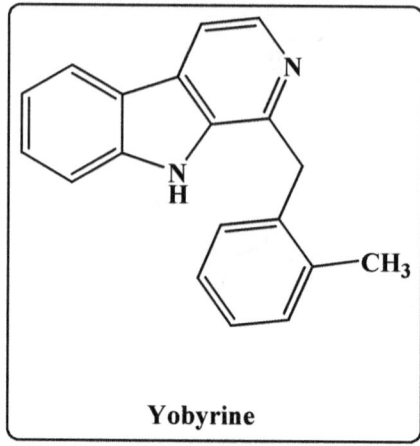

Yobyrine

K Synthesis of Yobyrine:

Yobyrine

v As Yobyrine is formed from reserpic acid, it means that reserpic acid may possess the following types of skeleton structures.

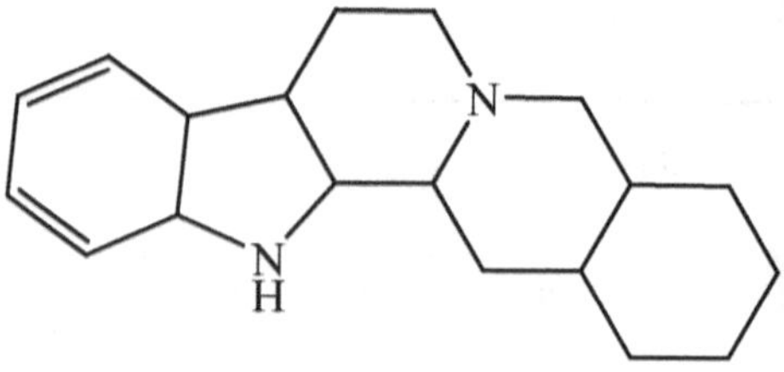

Skeleton Structure

vi From the fact, it conclude that one of the methoxy group is present in *m*-position to the >NH group of indole, it means -OMe group is present at C-11.

vii Reserpic acid when dehydrogenated, yields 11-hydroxy-16-methyl yobyrine, these may be only formed if –COOH group is present on C-16

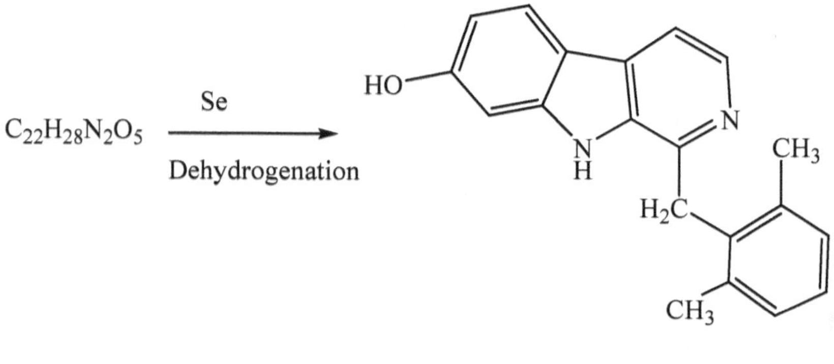

$$C_{22}H_{28}N_2O_5 \xrightarrow[\text{Dehydrogenation}]{\text{Se}}$$

11-hydroxy-16-methyl yobyrine

From the above discussion, It is confirmed that –COOH and -OH group are in *m*-position to each other, but –COOH group is present at C-16. Therefore –OH group must be present at C-18 position.

From purely biogenic reasons, the 2nd methoxyl group has been assigned position C-17.

On the basis above mentioned facts, the structure of reserpic acid may be as follow:

Reserpic acid

6 Structure of Reserpine:

As reserpine is di-ester of reserpic acid, it means that,

Resperine

7 Synthesis of Reserpine:

The structure of reserpine has been confirmed by its synthesis given by Woodward et.al.

cyclohexa-2,5-diene-1,4-dione buta-1,3-diene

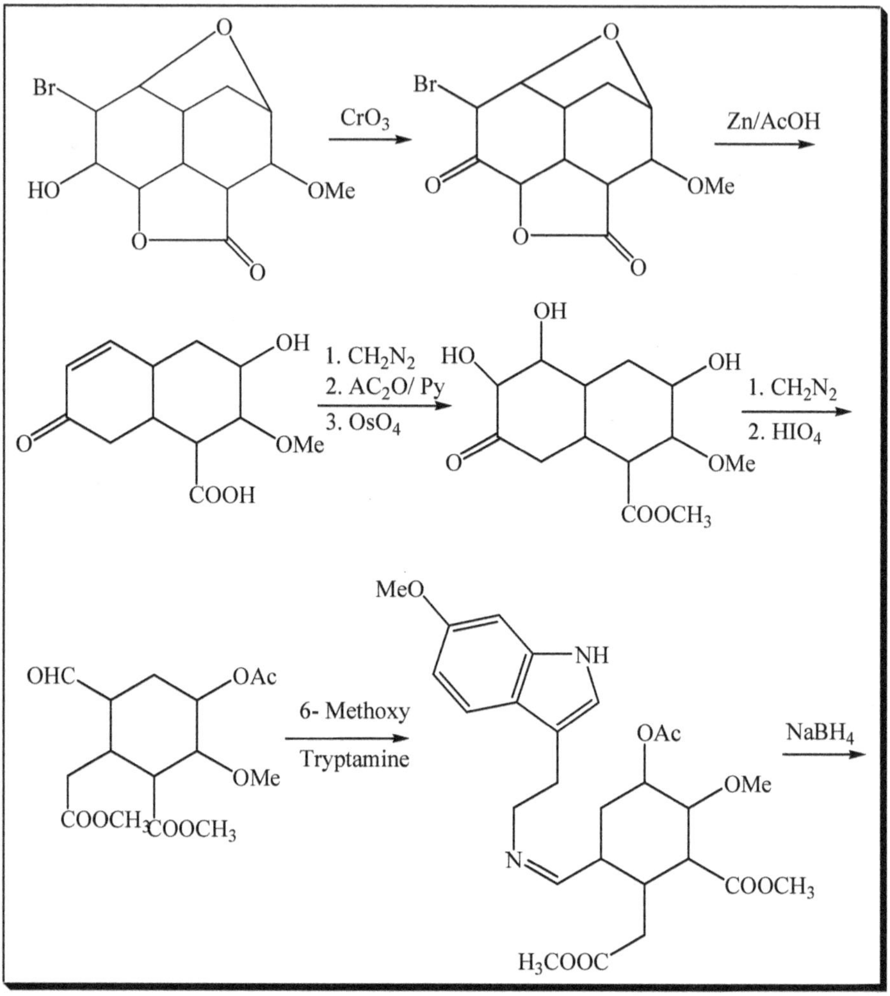

Br
HO
O
O
O
OMe
CrO3
Br
O
O
O
O
OMe
Zn/AcOH
OH
O
OMe
COOH
1. CH2N2
2. AC2O/ Py
3. OsO4
OH
HO
OH
O
OMe
COOCH3
1. CH2N2
2. HIO4
OHC
OAc
OMe
COOCH3
COOCH3
6- Methoxy
Tryptamine
MeO
NH
OAc
OMe
N
COOCH3
H3COOC
NaBH4

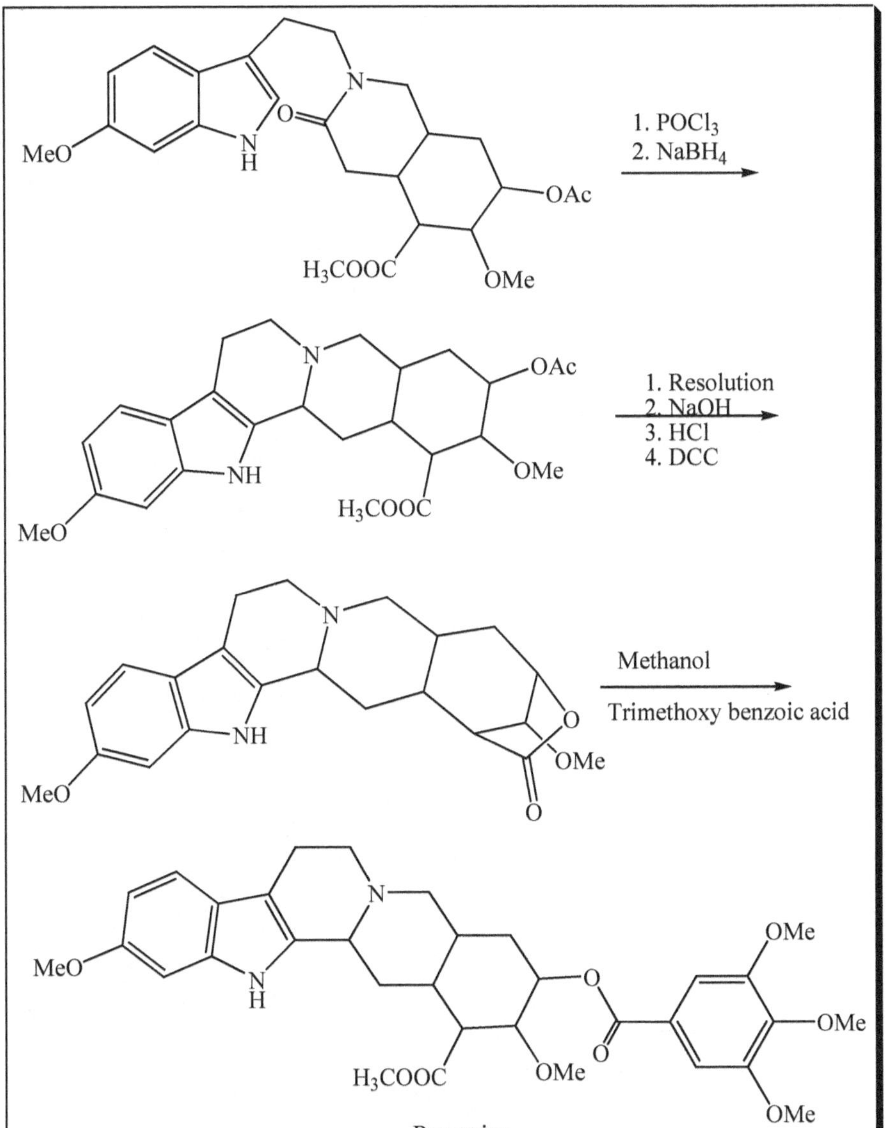
MeO
N
H
N
O
OAc
H3COOC
OMe
1. POCl3
2. NaBH4
N
OAc
NH
OMe
H3COOC
MeO
1. Resolution
2. NaOH
3. HCl
4. DCC
N
NH
O
OMe
O
MeO
Methanol
Trimethoxy benzoic acid
N
MeO
N
H
O
O
OMe
OMe
OMe
H3COOC
OMe

Reserpine

www.ingramcontent.com/pod-product-compliance
Lightning Source LLC
Chambersburg PA
CBHW021901170526
45157CB00005B/1919